OFF TO SPACE!

THE SUN

by Rachel Rose

Consultant: Beth Gambro
Reading Specialist, Yorkville, Illinois

BEARPORT PUBLISHING

Minneapolis, Minnesota

Teaching Tips

Before Reading

- Look at the cover of the book. Discuss the picture and the title.
- Ask readers to brainstorm a list of what they already know about the sun. What can they expect to see in this book?
- Go on a picture walk, looking through the pictures to discuss vocabulary and make predictions about the text.

During Reading

- Read for purpose. Encourage readers to think about the sun and what role it plays with Earth as they are reading.
- If readers encounter an unknown word, ask them to look at the sounds in the word. Then, ask them to look at the rest of the page. Are there any clues to help them understand?

After Reading

- Encourage readers to pick a buddy and reread the book together.
- Name three things the sun does for Earth. Find pages that tell about these things.
- Ask readers to write a sentence and draw a picture of something that they learned about the sun.

Credits:
Cover and title page, © Triff/Shutterstock; 5, © imtmphoto/Shutterstock; 6–7, © phive/Shutterstock and © egal/iStock; 8, © Amanda Carden/Shutterstock; 10–11, XXXX; 13, © Vladimir Arndt/Shutterstock; 14–15, © Pitris/iStock; 16–17, © Smileus/Dreamstime; 18–19, © Amanda Carden/Shutterstock, © guteksk7/Shutterstock, and © NPeter/Shutterstock; 20, © RyanJLane/iStock; 22, © Triff/Shutterstock; 23BL, © Orla/Shutterstock; 23BM, © adventtr/iStock; 23BR, © NPeter/Shutterstock; 23TL, © Xurzon/iStock; 23TM, © SERDTHONGCHAI/Shutterstock; 23TR, © Art-Perfect/Shutterstock

Library of Congress Cataloging-in-Publication Data

Names: Rose, Rachel, 1968- author.
Title: The sun / by Rachel Rose ; consultant, Beth Gambro, Reading Specialist.
Description: Minneapolis, Minnesota : Bearport Publishing Company, [2021] | Series: Off to space! | Includes bibliographical references and index.
Identifiers: LCCN 2020042958 (print) | LCCN 2020042959 (ebook) | ISBN 9781647475093 (library binding) | ISBN 9781647475147 (paperback) | ISBN 9781647475192 (ebook)
Subjects: LCSH: Sun—Juvenile literature.
Classification: LCC QB521.5 .R67 2021 (print) | LCC QB521.5 (ebook) | DDC 523.7—dc23
LC record available at https://lccn.loc.gov/2020042958
LC ebook record available at https://lccn.loc.gov/2020042959

For more information, write to Bearport Publishing, 5357 Penn Avenue South, Minneapolis, MN 55419.

Contents

Good Morning!

The day starts when the sun comes up.

It gives us light.

Time to wake up!

The sun is a star.

It is the closest star to Earth.

But it would still take 19 years to fly there by plane!

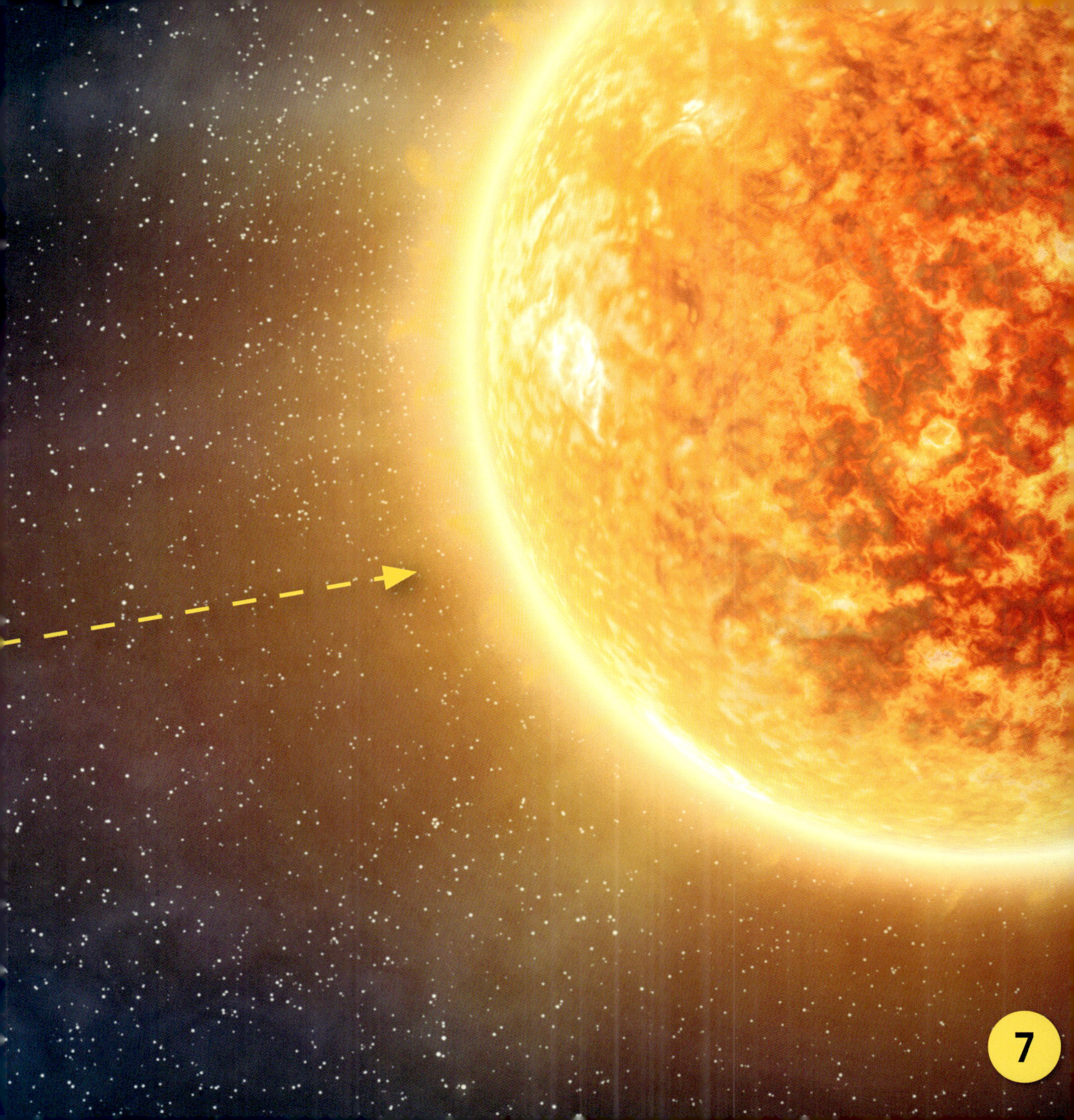

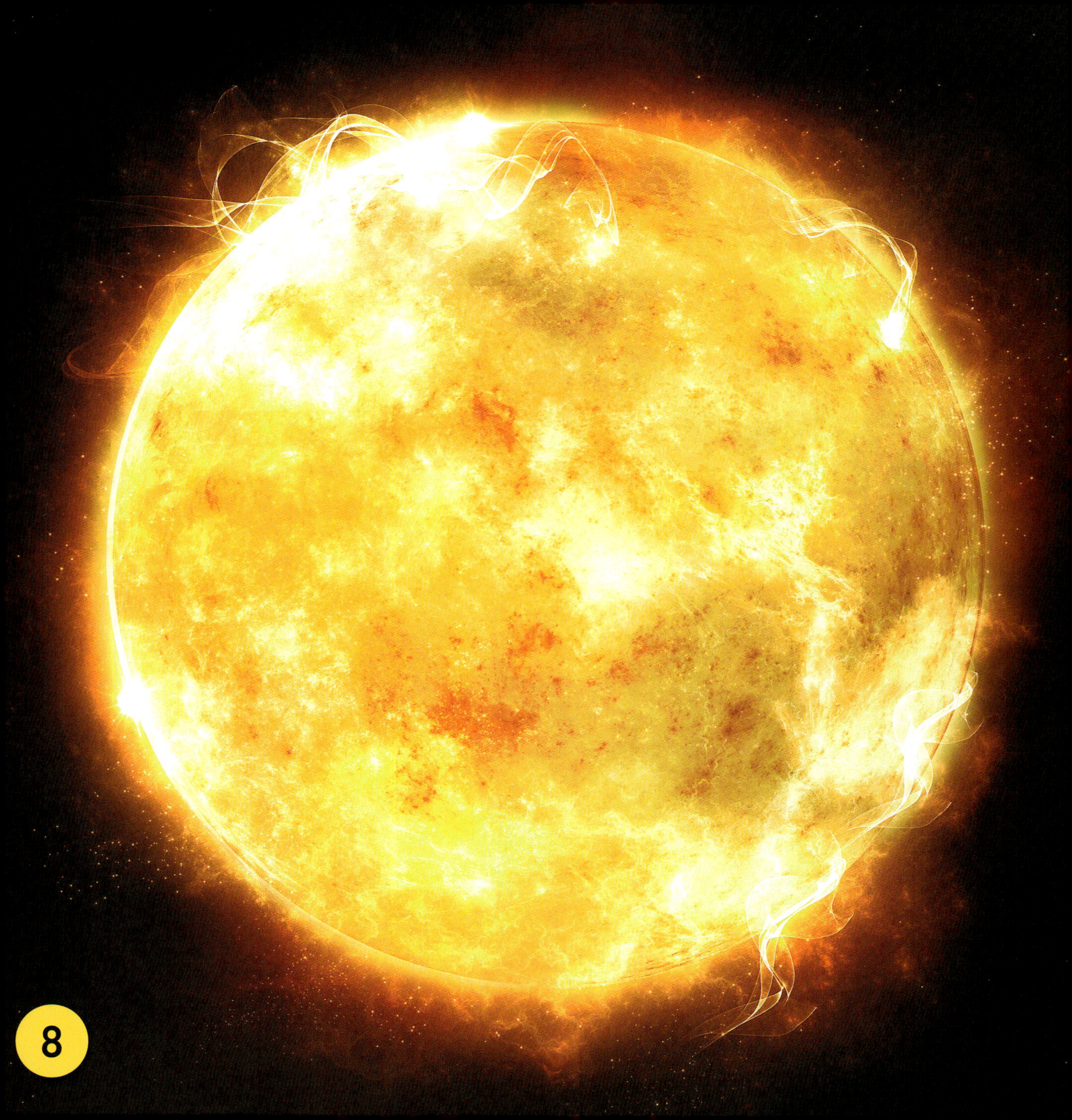

Like all stars, the sun is a big ball of **gas**.

It has a lot of **energy**.

This makes it very hot and **bright**.

The sun is the middle of our **solar system**.

Earth circles the sun in a path called an **orbit**.

Other planets do, too.

The sun
Earth

The sun may not look very big.

But it is huge.

You could fit more than one million Earths inside it!

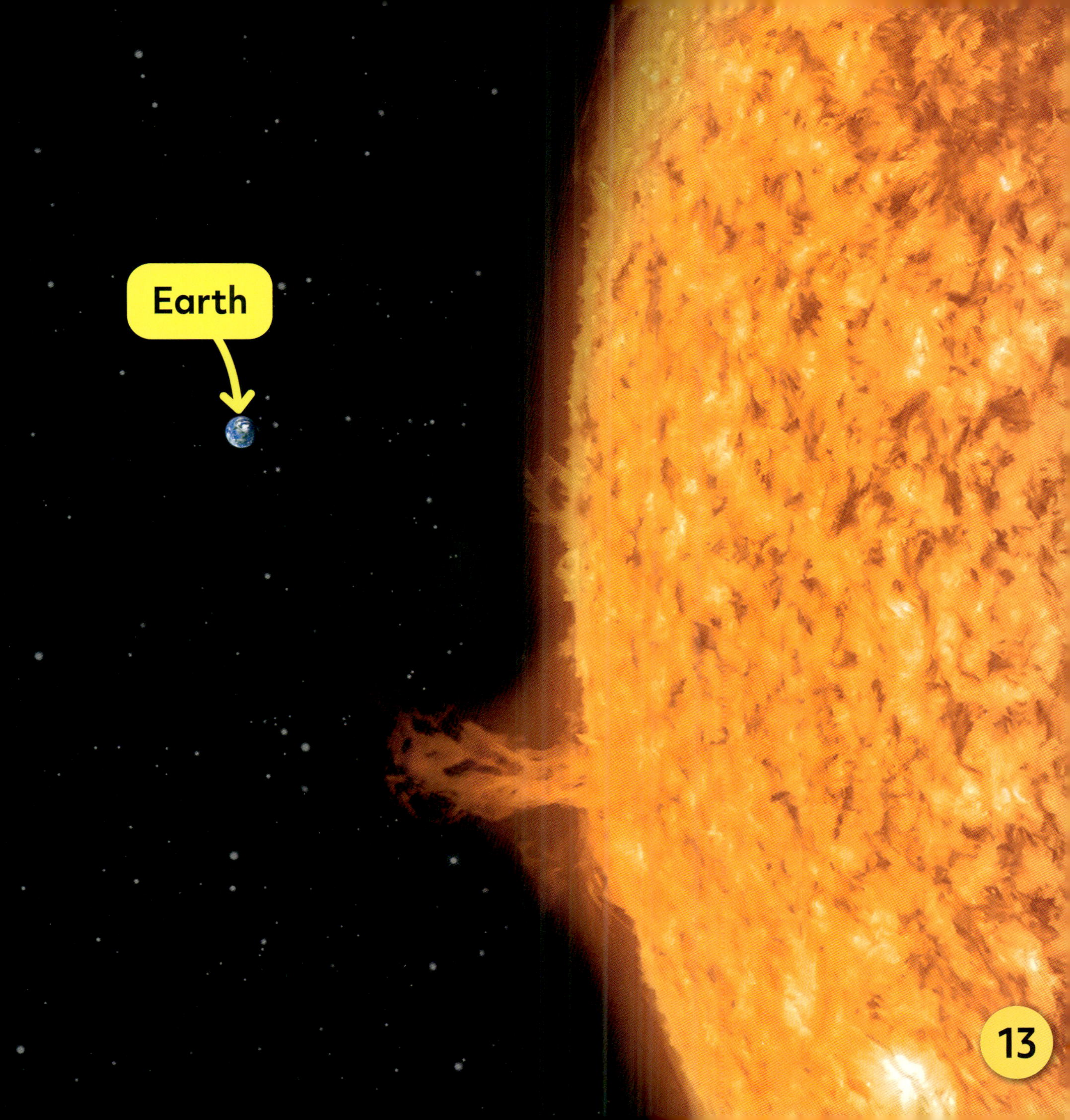
Earth

Night

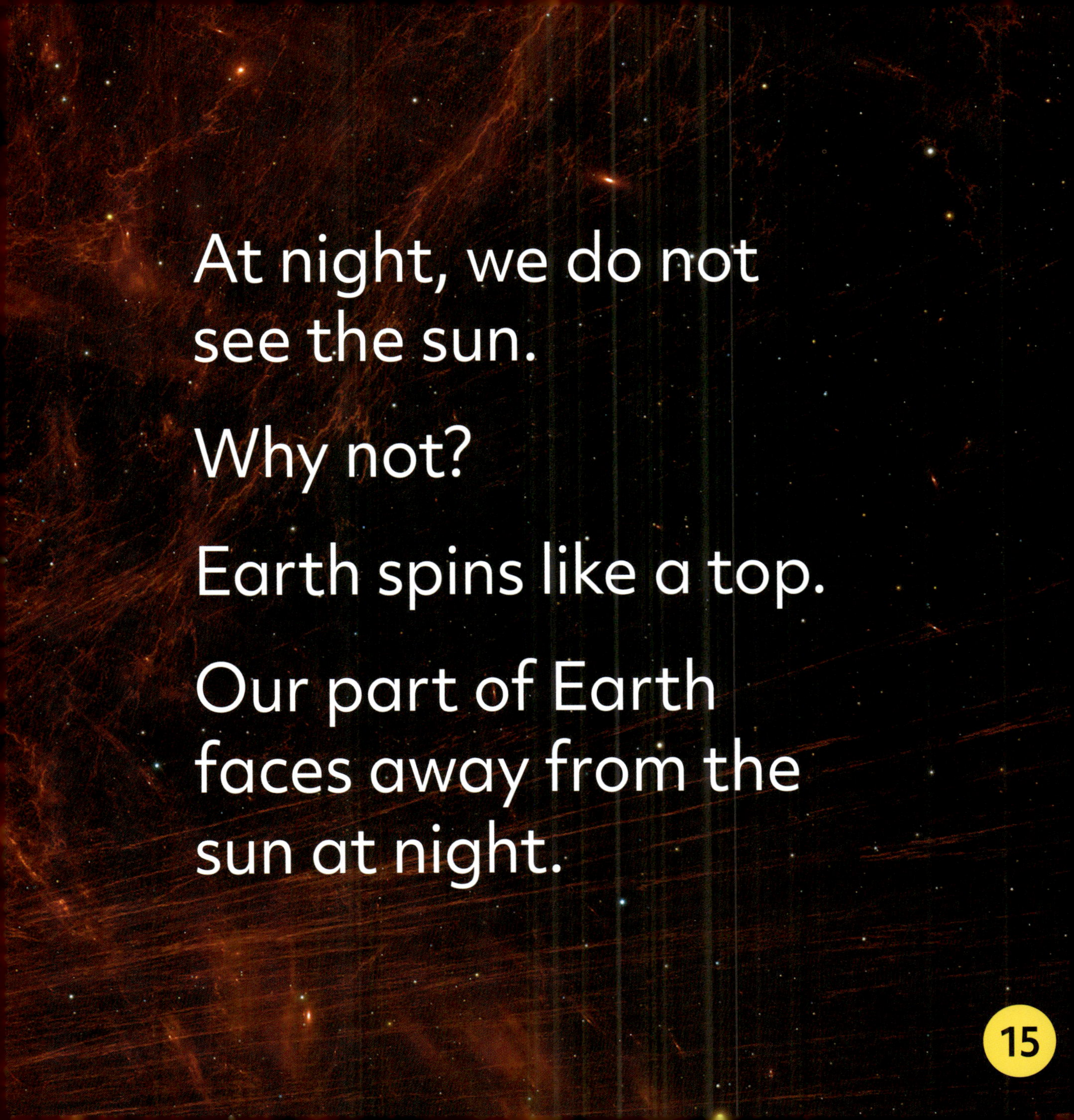

At night, we do not see the sun.

Why not?

Earth spins like a top.

Our part of Earth faces away from the sun at night.

During the day, our part of Earth faces the sun.

This is when we see the sun's light.

We have seasons because of the sun, too.

Earth **tilts**, or tips, in space.

The part that tilts to the sun has summer.

The part that tilts away has winter.

Seasons change as we orbit the sun.

We could not live without the sun.

It brings us heat and light.

It gives us energy to grow.

Thank you, sun!

Fun Sun Facts

It takes eight minutes for the sun's light to get to Earth.

The sun is the only star we can see during the day.

The sun is white. It only looks yellow from Earth.

Glossary

bright giving off strong light that is easy to see

energy power that makes things work or grow

gas something like air, with no shape

orbit the path of an object that is circling a planet or the sun

solar system the sun and everything that circles around it

tilts leans to the side

Index

Read More

Hansen, Grace. *The Sun (Our Galaxy).* Minneapolis: Abdo Kids, 2018.

Moon, Walt K. *Let's Explore the Sun (A First Look at Space).* Minneapolis: Lerner Publications, 2018.

Learn More Online

1. Go to **www.factsurfer.com**
2. Enter "**Space Sun**" into the search box.
3. Click on the cover of this book to see a list of websites.

About the Author

Rachel Rose lives in California.
The sun shines there a lot!